CLIMATE CHANGE REALITY CHECK

BASIC FACTS

THAT QUICKLY PROVE

THE GLOBAL WARMING CRUSADE

IS WRONG

<u>AND</u> DANGEROUS

CALVIN FRAY

ISBN: 1530535352
ISBN-13: 978-1530535354

DEDICATION

To Mom and Dad, who taught me the difference between right and wrong, the importance of acting on it, and the value of always leaving a place at least as good or better than I found it. I hope this book makes you proud.

ACKNOWLEDGMENTS

To all of those whose passionate commitment to truth, reason, and the freedoms to think, speak, and live up to our fullest potential have served as courageous examples. Thank you for having a mind and having the self-confidence to express it. Then act on it.

"What is popular is not always right.

What is right is not always popular."

-- Anonymous

LAYING IT OUT THERE

Let's be straight up and honest with each other. I'm not a climatologist or meteorologist. But I *am* someone who cares deeply about truth, science, reason, politics, economics, and the world in which we all live. Beyond that, I care passionately about living in a world in which as many people as possible can realize their fullest potential as individual human beings.

If you can relate to this and sympathize with it, then we have more to talk about.

My technical and professional background is engineering and project management. I have a Bachelor's and Master's degree in aerospace engineering, plus a professional certification in project management.

I'm a very analytical person and love solving problems. But before getting to that point—I love taking the time to think and study a situation so that I know exactly WHAT PROBLEM needs to be solved.

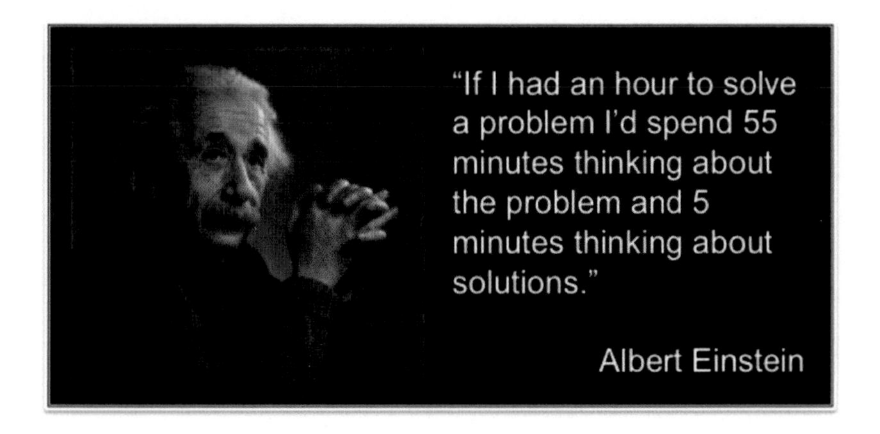

There was a great role model!

Given that context, and my persistent desire to apply my analytical reasoning skills to **get to the point** of an issue so that we can UNDERSTAND THE REAL TRUTH of something, so that we can then DO WHAT REALLY MATTERS and what HAS THE BIGGEST IMPACT, here is my take on the **anthropogenic global warming (AGW) controversy.** It has also come to be known as **the great climate change debate.**

This analysis is going to take you less than one hour to read. Probably closer to 30 minutes. Within it, you'll get to the root of the issue that will take you less than 5 minutes with someone else to prove that the current so-called "consensus" is quite insane, to put it bluntly. Not only that, but it is inherently and fundamentally dangerous itself to the future prosperity and freedom of our species.

After reading this book, I'm confident that you will not look at this issue the same way again. And you will have a better understanding of the physics and science of AGW than 99.9% of the population.

If you want to earn easy money from a bet against an AGW climate change believer, with facts that you can quickly prove with a simple check from Wikipedia or any science reference textbook, you'll get that too.

Sometimes the most serious topics require a little cheekiness and sense of humor, just to keep us sane and grounded. You'll get some of that in this book too.

WHAT THIS BOOK IS NOT ABOUT

This book is not about the solutions. It is not about consequences and effects of global warming or climate change.

That's what infuriates me so much about so much of the literature, discussions, and propaganda about climate change! It already assumes it is going to happen!

So many people and "arguments" focus on the scary, dramatic impacts of what could or would happen IF the predictions come true.

That is NOT PROOF that it is going to happen! And it is NOT AN EXPLANATION about the CAUSE behind it! (Instead, they are various types of *non-sequiturs,* logical fallacies, or irrelevant distractions masquerading as rational discussion.)

Where is the book or article that explains the basic science about how and why this is such a big problem in the first place? How did we get from the hole in the ozone layer…to the greenhouse effect…to global warming…then to climate change…then to CO_2 emissions…then to the need for Carbon credits?

You are about to learn how all of those are related…*or not!*

If you or someone you know is so passionate about limiting the growth of CO_2 gasses, shouldn't we have a basic understanding of how that relates to climate change?

Yes! Especially if there are political policies and regulations that are going to impact our daily lives. That is exactly what is in motion across the planet, in case you haven't been paying attention. But since you are reading this book, I presume you've noticed.

This book is not about the details and nuances of climate change models. It is not about any theoretical physics. It is not about complex climatology models.

All of those things exist. But to be valid, they have to be based on the fundamental physics and laws of nature that we are all governed by.

The basic physics is something that we can all understand. If you have had any science class at any point in your life, I'm going to keep it to that level. If you have a high school diploma you are smart and educated enough to follow this analysis.

But first, if we want to take a rational, logical, and practical approach to the problem, let's try to agree on some basic fundamentals. This is where I'm coming from. If you don't follow along with me, I won't expect you to agree with my conclusions. But if you also value a reality-based opinion and rational approach, this book captures the conversation I've had with myself.

FUNDAMENTAL TRUTHS

First, there is an objective reality and truth about what is happening with our climate. Whatever is happening, it is real. And it doesn't matter what we as individuals or as a species believe or think or wish about it.

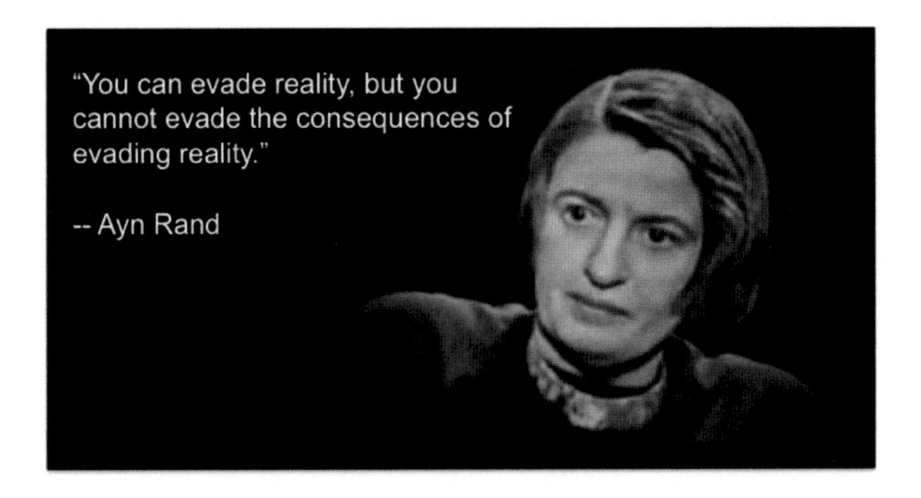

Hopefully that isn't a point of controversy or argument with you, dear reader. But there are plenty of people in the world who disagree with that basic premise. If someone believes that we can wish or meditate or pray the problem away, through pure consciousness, we're already on a different page. Different planet and universe, as far as I'm concerned!

Second, whatever the facts truly are, we are capable of understanding it. Probably not with 100% certainty and 100% granularity, but by using our powers of perception, analysis, technology, and rational discussion, we can find the truth together.

The situation is *not* inherently unknowable or hopeless.

Hopefully you agree with that premise too. It's important to believe in ourselves and our ability to understand the world in which we live. Whether you believe we have evolved to be uniquely and beautifully suited to survive and adapt in this world over millions of years, or we have been supernaturally designed to live and prosper on this Earth, you can share this premise.

Third, I believe we as a species have the power and ability to find a peaceful, feasible, and sustainable solution. We have learned how to not only live, but also prosper in the harshest of environments across the planet. We have even lived beyond our planet!! Admittedly, we haven't advanced far beyond the temporary survival phase in outer space. The same is true underwater. We haven't learned how to sustain ourselves in either environment without help from the Earth-bound surface dwellers. Yet!

But let's remember the context of our discussion. With **creativity, resourcefulness, innovation, and technology**, we can certainly deal with a few degrees (or fractions of a degree) of temperature change. Right?! We can deal with a few inches (even feet) of water level change over decades or centuries.

CONTEXT MATTERS

We can't forget the CONTEXT of what we are investigating and talking about! We aren't talking about our sun going supernova or swallowing our planet as a red giant anytime soon…we still have a few billion years to prepare for that!

Which brings me to one more important point: **PROPORTIONALITY**. This is an essential principle of successful engineering and design thinking. Pareto is famous for the concept of the Pareto Principle, which is also known as the 80/20 rule. If climate change is a genuine threat to our species, *what are the highest impact or highest leverage actions we can take?*

Chances are, there are a few high-leverage actions we could take (20%) that would produce most of the results we seek (80% impact). Shouldn't we look for these high-leverage actions?

I'm a very practical and goal-oriented person. If there is a genuine, urgent, and major problem, let's not f__ around about it. Time, money, and people's lives are too precious to waste pursuing things that don't matter or are ineffective or wasteful.

By the way, don't fall for the sloppy thinking or hyperbole that OUR WORLD IS THREATENED BY CLIMATE CHANGE! "*Our world" means our planet Earth.*

Earth itself is not threatened!! Get a grip! Remember the context!! The point we must remember–*the context that really matters*–is the effect (if any) of these changes on HUMAN LIFE.

HUMAN LIFE must be the standard by which we evaluate the <u>impact</u>, the <u>risk</u>, and the <u>effectiveness</u> of any actions we take.

I hope you are still with me. Just to make my point with an argument to the point of absurdity–do you know one **100% guaranteed sure-fire way** to eliminate the threat of human-caused global warming?

Get rid of all the humans! Done. Now we can all go home happy and celebrate. Except none of us will be here to enjoy it…

Don't laugh too hard, comrade. If you favor a solution that is fundamentally incompatible with human life or human prosperity, you are not doing yourself (or the rest of us) any favors. In fact, you are part of the problem, not the solution.

This is the reason why so much of the global warming crusade is fundamentally dangerous—because the proposed "solutions" are a bigger threat to human life and well-being NOW than any potential climate change decades in the future.

Please think about that before you advocate stripping away our ability to innovate! Or talk about discouraging our ability to develop more efficient and advanced technologies in order to make better use of the resources we have available.

Those resources include the **brainpower and ambitions of scientists, engineers, entrepreneurs, and innovators from all walks of life.**

FACTS VS. ARGUMENTS

Now let's get to the facts. How many of us actually understand the science and physics behind climate change? With all of the hype and fervor and activism around the topic, why are so many people relying on **arguments from authority**?

A clear consensus of experts agree…

Or:

All of the *honest* experts agree…

Or the most popular one:

There is a clear consensus among scientists…

That argument has two flaws: the argument from authority, and the argument from consensus (or peer pressure).

For something as important as climate change, which is claimed to be such a large and urgent threat that it threatens the very existence of our species, **shouldn't every one of us understand the fundamentals of what is really happening?**

Yes!

Definitely, those who are advocating for taking action or changing our current lives in some way should know what the hell they are talking about. Perhaps that means you…

THE SCIENTIFIC METHOD TO THE RESCUE

I'm going use my powers of engineering, analysis, and reason to get to the truth and share it with you right now. I hope you are ready for it.

Let's use the scientific method to approach this problem. Again, I hope you agree this approach works and is a smart way to approach our situation.

What is the main hypothesis or claim? It is this:

Human activity is causing global climate change (specifically, global warming)

Remember the importance of focusing on the highest impact causes and effects (the Pareto Principle). What claimed to be the primary driver of anthropogenic global warming?

The answer we hear is:

Carbon dioxide output (as a greenhouse gas) is the primary driver of anthropogenic global warming.

Now, this has evolved or grown into talk about Carbon itself... Think about how many times you've heard proposals about creating schemes and markets for CARBON TAX CREDITS or

CARBON CREDITS.

For now, let me just remind you that CARBON IS THE FUNDAMENTAL BUILDING BLOCK OF LIFE on our planet. So, if CARBON is claimed to be the root cause of our problem with global climate change…where do you think that leads us?

What if there was a tax based on carbon? Guess what, my friend. You've just created a scheme to **justify a tax on <u>every living being</u> on the planet.** Not just humans…I wonder how we will collect taxes from all the other species being so greedy (!!) with all of those carbon atoms…

Anyway, let's focus on CO_2 (Carbon Dioxide).

Here's a question I love to ask every person who believes CO_2 is the primary culprit behind climate change. When I do this in person, I often make it a bet for real money.

HOW MUCH OF OUR ATMOSPHERE IS CARBON DIOXIDE?

If I offered you money to answer correctly to the nearest 1%, could you do it?

Waiting…

What's your answer?

Let's go back to our Pareto Principle, shall we? I hope this doesn't come across as condescending. But since so many people seem to lack basic scientific knowledge about our planet, and a consistent inability of applying the scientific method, we will find it very helpful to take the time and effort to go through this process.

Do you remember from school **what is the most common element in our atmosphere?**

Think with me…

Hopefully you said NITROGEN.

And maybe you recall; it makes up a whopping **80% OF OUR ATMOSPHERE.**

It's actually 78%...but how much is 2% worth when you are talking about our entire global atmosphere? (That's a sarcastic joke…and it will get funnier in a minute.)

Obviously, there isn't any hysteria about Nitrogen…

Let's move to #2… What is the **second most prominent element in our atmosphere?**

Hopefully you know the correct answer is: OXYGEN.

Hopefully you also know that Oxygen is what we humans (and almost all animal species) need to live and breath.

Do you remember or know how much of our atmosphere is Oxygen?

The answer is…

approximately **20%.** (It's actually closer to 21%.)

What the…!?!

There's that Pareto Principle of the 80/20 rule showing up again!!!

No, that's not what is so shocking. :-) If you add 78% + 21% you get…

99% !!!!

WTH?!? Where is CO_2 in this equation?!

It must be #3, right?

No, sorry. It turns out that the #3 spot goes to **Argon, at 0.93%.** That might as well be considered 1%.

WTF?!? ***Now we are at 100% and there is still no Carbon Dioxide!!!***

Check the facts yourself…it turns out that the amount of CO_2 in our atmosphere, to the nearest percentage, is:

ZERO!

More exactly, it is <u>0.04%</u>. And that is rounded off slightly. More precisely:

0.039% of our atmosphere is carbon dioxide

Don't believe me? Go to Wikipedia or your favorite trusted resource yourself. In fact, I don't want you to take my word for it. Use your own mind and judgment–please!

Remember the importance of PROPORTIONALITY?

If you believed that CO_2 was **1% of our atmosphere,** you would be too high off the mark by a <u>complete order of magnitude</u>.

If you thought that CO_2 was just **0.1% of our atmosphere,** you would still be too high by an entire decimal place.

Rounding off some numbers with +/-2% accuracy seems very sloppy and careless now, doesn't it?!

But we are just getting started…don't leave me yet.

HOW TO EARN EASY MONEY KNOWING THE TRUTH ABOUT CLIMATE CHANGE

Remember that hypothetical bet I offered about correctly answering the amount of CO_2 in our atmosphere to the nearest full percentage? I have yet to find someone who gives **the correct answer of ZERO.**

There is no other way to round off 0.039%!!!

Now you have a fun way to earn some easy money the next time you are with friends, family, or climate change enthusiasts.

Here's how a conversation might go:

AGW Believer (AGWB): We need to do something about this climate change threat…(or some statement that implies this).

YOU: I'm still trying to understand the reason for the concern, because I don't get it. What's your understanding of the biggest culprit or cause of global warming?

AGWB: Humans are causing it, of course!

YOU: OK, and how exactly? What is the one thing we are doing that is blamed as the key factor?

AGWB: well, it's carbon dioxide (or CO_2) emissions

YOU: Right, that's what I hear too. But if CO_2 is such a big factor, it seems like it should be a significant part of our atmosphere, right?

AGWB: Well, I guess so.

YOU: That's what I thought too. Then I found out the truth. I looked it up because it's a simple fact of nature on our planet. But nobody is considering or talking about it for some reason. And when I learned the truth, it called everything into question that we are being told.

In fact, I'll be you $___ (pick a number you are comfortable with. I like $10.) that you can't tell me what percent of our atmosphere is CO_2. It's a small number, so I'll give you to the nearest 1%. Wanna bet? ☺

Where it goes from there will get interesting…but this is a very valid and effective question to even ask the people involved in climate science!

WHAT TO ASK A REAL SCIENTIST OR RESEARCHER

Ask them the accuracy of their computer or simulation models. You'll get a number like 10-20% if they are being generous. More likely, they won't give you any number. "It depends" is actually a reasonable and honest response. Are you asking about temperature? At what location and point in time? General trends?

Ask them to quantify whatever type and amount of accuracy they want. Let them define the assumptions and parameters! If they are willing to give an accuracy of 50-100% for anything you might have found an honest researcher.

I've written and used many complicated computer simulation models in my career. They were for fluid and thermal dynamics simulations of automobiles and airplanes. These programs took several minutes to several hours to run on modern computers. The accuracy of these was on the order of 1-5%. The results looked something like this:

That level of accuracy is very good—it's a small number. The accuracy was based on running an actual experimental test of a comparable vehicle in a wind tunnel, with extensive instrumentation throughout the vehicle and in the facility.

That is very different from running a simulation of the *entire planet's atmosphere!*

And it would have been foolish—impossible—for those computer simulations to try and track the influence of a molecule that was 0.039% of the air volume!

The resolution of these two systems is simply too far apart! They are fundamentally mismatched. It would be like trying to measure the dimensions of your living room with your car's odometer.

You laugh—but it's a valid analogy! A car's odometer measures distance on the scale of miles or kilometers. The first decimal is in tenths, and then you can't tell anything more detailed than that. (A tenth of a mile is 528 feet.)

A typical room is 10 to 50 feet in length, if it's large. Best case, you would have to drive through <u>ten rooms</u> before your odometer turns one number over in the tenths digit.

Basic fact of science and engineering:

You can't accurately measure or analyze (never mind PREDICT) anything that is smaller than your overall level of accuracy!!

How many times larger would a material that is 0.04% of a system have to be when your system accuracy is 4%??

Answer: 100 times greater!

So actually, my analogy was far too generous. If an odometer resolution is 528 feet (more or less), you would have to use your car to try and measure the length of something that is about 5 feet long.

Anyone want to lie on the ground and try and get their height measured with a car's odometer? Puh-lease! That is obviously silly.

I'm using that analogy to make a point. Measuring the effects of CO_2 in a global climate model isn't exactly like that. But more on analogies later...

But we aren't done looking at CO_2 yet. It is getting so much attention in the climate change crusade that we need to keep going.

HOW MUCH OF THE CARBON DIOXIDE IN THE ATMOSPHERE IS FROM HUMAN ACTIVITY?

That's what we need to worry about, right? And that's what 'all of the experts' are saying we need to take drastic action to reduce.

But what do the actual numbers tell us?

As summarized by www.skepticalscience.com (but don't trust me–do your own research), there are **730 gigatons (GT) (also billion tons) of CO_2 in the atmosphere.** (BTW, that website is built by AGW proponents.)

Humans release 29 gigatons of CO_2 into the atmosphere per year. (From what I can find from my research. Again, *YOU* should do the work to find the number and convince yourself if you really care about the truth.)

But **the amount of CO_2 that is increasing in atmosphere is 12 gigatons per year.**

What do you know, it is actually increasing!

Let's acknowledge and admit when we learn something that challenges our assumptions and beliefs. I'll even emphasize it for us both:

The amount of CO_2 is increasing by 12 gigatons each year. This is 1.6% of the existing atmospheric CO_2.

Now, let's look at the actual numbers to get a sense of the

proportions.

The net load from human activity is 12 gigatons per year. To say another way, <u>17 gigatons of what we create (29 – 12) is getting consumed (offset) by other natural forces.</u>

The largest of these is PLANT LIFE, which have the opposite relationship of Oxygen to Carbon Dioxide via photosynthesis. (For those that need a biology refresher, plants ingest carbon dioxide and expel oxygen. We humans, like most animals, do the opposite.)

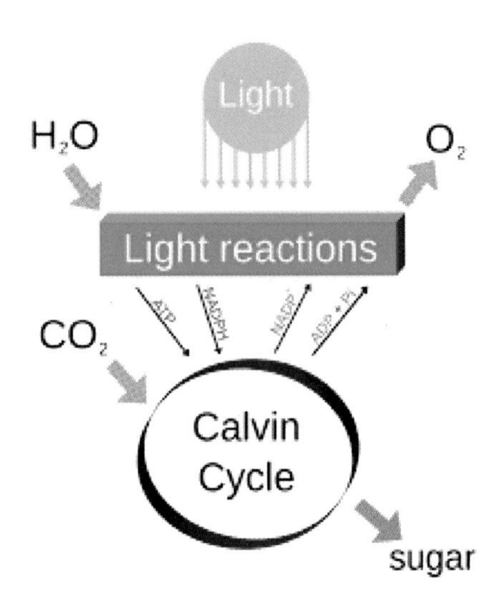

Photosynthesis

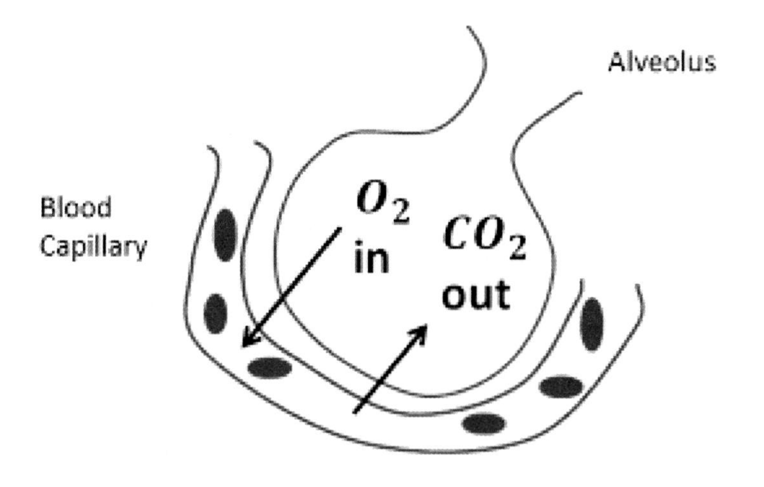

Respiration

If human activity reduced our CO_2 output by 12 GT/year, the amount of CO_2 in the atmosphere would stabilize. (As best we can measure and understand the global climate mechanisms today, at least.)

The amount of reduction for us would be very significant: 12 / 29 is **a 41% reduction.** Try reducing your budget (of energy or money) by 41% and see how well you can maintain your standard of living. You probably can't!

But here is the bigger picture.

How does the TOTAL HUMAN OUTPUT of CO_2 compare to the TOTAL QUANTITY OF CO_2 IN THE ATMOSPHERE?

That answer is 29 / 730, or

4.0%

More exactly, it's 3.97%. To put that in another perspective, **96% of CO_2 in the atmosphere is going to be there no matter what humans do**.

Keep in mind; CO_2 is 0.039% of the atmosphere anyway. So all of this research and frenzy is about the human-created portion (4.0%) of 0.039% of the atmosphere. What is 4.0% of 0.039%? We are talking about **0.0015% of the atmosphere!**

0.0015% !!

How much of your tax dollars or your politicians' attention do you think should be spent on a problem of that size?

LOOKING AT THE GREENHOUSE EFFECT

But let's look more into these greenhouse gasses. We aren't done yet! Maybe CO_2 is actually a big factor, even though it's essentially a trace gas.

Do you know what the most powerful greenhouse gas is? Again, let's use the Pareto Principle as a guide.

It isn't CO_2. It's something even more familiar and precious to all of us...

WATER.

It's also known as H_2O.

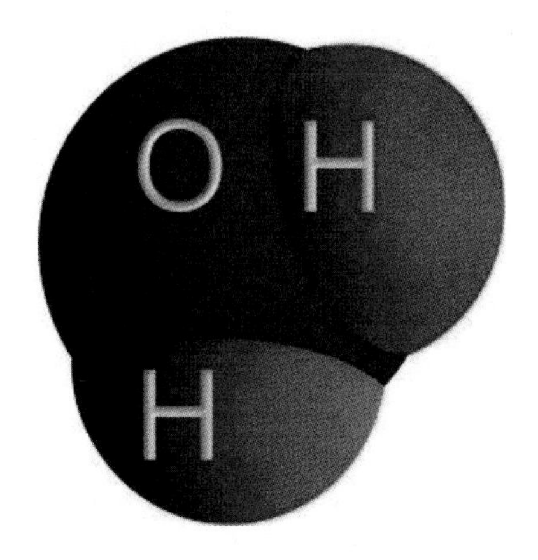

How is that, you may ask?

Greenhouse gasses trap heat created at the Earth's surface and prevent that heat from escaping into outer space. This phenomenon is critical in the climate change debate, so you need to understand it. And you are about to.

What is standing in the way between the Earth's surface (land or water, doesn't matter) and outer space? It is our atmosphere, of course.

This is where the physics gets more complex…so let's use an analogy. Our atmosphere is like a big insulation blanket. It dampens or filters out some energy from the sun as it comes in to Earth (like some types of radiation). And it also acts as an insulation blanket in the reverse direction – keeping heat IN after the land or water has been heated by the sun's energy (or more heat is created from Earth itself, such as volcanoes, forest fires, life, etc.)

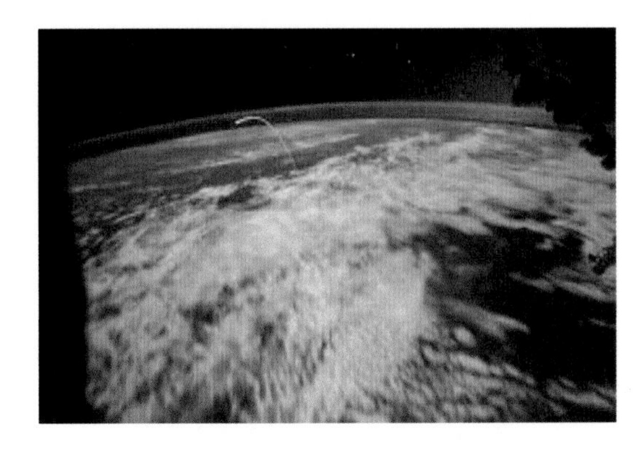

Please keep in mind—an insulation blanket is an ANALOGY. It isn't a perfect or absolute explanation.

Just like the greenhouse effect is also an analogy, let me point out. We are not actually living in a greenhouse, with clear plastic sheets above us.

As far as analogies for our atmosphere go, a blanket seems like a much better one than a greenhouse. But that's just me.

Anyway... Let's get into the real science that matters!

THE FUNDAMENTAL PHYSICS PRINCIPLE BEHIND GLOBAL WARMING

The scientific term to appropriately measure and compare this insulation capability for gasses is HEAT CAPACITY. The Engineering Tool Box website has a list of heat capacities for some common gases.

Heat capacity measures how much a temperature of a material will change when it is exposed to a unit of energy. It is a function of the element or compound. It is also a function of how dense or concentrated that material is.

To account for the material density, we need to decide if we are going to compare these gasses at a constant pressure or a constant volume.

We want to use the value at a constant volume, because air pressure changes with altitude. Another way to justify this decision is to say that when you consider the total of Earth's atmosphere, it is a fixed volume (between the surface and the edge of space). The values of heat capacity for constant volume are expressed as c_v, and are in metric units of kJ/(kg K).

Gas	**c_v (kJ/(kg K))**
Air (combined)	0.718
N_2	0.743
O_2	0.659
CO_2	0.655
H_2O vapor	1.46

First sanity check–do you notice how the heat capacity of air is in between the values of nitrogen and oxygen? And it's closer to the value of nitrogen. That makes sense, considering that N_2 is 78% of our atmosphere and O_2 is 21%.

Notice also how the heat capacity of CO_2 is nearly the same as O_2. And there is **OVER 500X more oxygen than carbon dioxide in our atmosphere**. Maybe we should look instead at reducing the amount of oxygen on the planet…ha ha.

But here's the more important point. **Water vapor has TWICE the heat capacity of air.** According to Wikipedia, the amount of water vapor at high altitudes (where the greenhouse effect is claimed to be most important) ranges from 0 to 4%. Let's assume a conservative average of 1%. I couldn't find an average value from any source, so if you have a better number I'll be glad to learn it and use it.

If the amount of water vapor in the atmosphere is 1%, which is MORE THAN 25 TIMES the amount of CO_2 (1 / 0.039 = 25.6).

To compare apples to apples, we need to compare the heat capacities of these two gasses, for an equivalent volume.

The proportional heat capacity of water vapor vs. CO_2 is 1.46 / 0.659 or 2.2x greater.

And we estimated (conservatively) that there is **25.6 times more water vapor than CO_2 at high altitude**.

That means that **water vapor is 2.2*25.6 = 57 TIMES more influential on the greenhouse effect than carbon dioxide.**

Forget oxygen...how about we start a global campaign to reduce the amount of water on the planet?

Let's get real, people! That would be ridiculous.

But if you really care about reversing the trend of global warming, it is WATER in the atmosphere--not carbon dioxide--you need to focus your activism on.

Or Oxygen… which would make it hard for anyone to get into a vigorous debate without being a hypocrite, with all of that heavy breathing! Come to think of it…the same goes for CO_2…hmmm.

Water vapor is what sometimes turns into clouds, which have the nice benefit of blocking the sun's rays and creating a shade underneath them. So less water in the atmosphere would also mean less clouds, which would seem to be a bad thing for global warming…the whole cloud mechanism in climate change models is still a big point of uncertainty and contention. It is complicated and complex…a big challenge to model confidently in computer models and simulations, for instance.

If the importance of something that is 57 times more powerful than carbon dioxide when it comes to heat absorption is still under debate in computer models, why are so many people so certain about the effects of 0.04% of the atmosphere?

To recap, does it make sense that a gas that is **only 0.04% of our atmosphere** is the driving force behind global warming and climate change?

Does it make sense that a gas with **1/57th the effective heat capacity of water vapor** is the driving element behind global warming?

How much of a country's GDP (or the entire planet's) should be spent on a phenomenon (human-created net increase in CO_2) that is **0.0015% of the atmosphere?**

I can only conclude that the most rational and proportionate response should be the answer that corresponds to the answer of all of these other questions:

ZERO. Nada. Zilch. Nothing.

WHAT IF HEAT CAPACITY IS A RED HERRING?

There will be some people who say that heat capacity is not the primary physics principle behind global warming and the greenhouse effect. If I can gain a better understanding of the situation, I'll be genuinely glad to.

But if it isn't heat capacity, what is it? Is it radiation and energy?

A MORE BALANCED LOOKED AT ENERGY BALANCE EQUATIONS

Many climate models are based on energy balance. Energy comes into the Earth from the Sun. Some energy flows back out into space. Then there are lots of things happening with energy on our planet and in the atmosphere. Additional sources of energy (natural and man-made), absorptions or "sinks" of energy (here also, natural and man-made), and reflectivity of energy back to the Earth that might otherwise leave the planet.

This last component is the famous greenhouse effect. This is where the radiation qualities of CO_2 come into play.

Here's a quick and simple reality check to ask yourself how important and significant the radiation (or heat energy) qualities of CO_2 can possibly be.

You can't see CO_2 in our atmosphere. Even if it had a color, spotting this 0.04% fraction of our air would be impossible. But look up into the sky, day or night, and what do you usually see?

CLOUDS!!

Do clouds have any impact on the amount of energy balance (heating and cooling) on our planet?

Yes, of course!! You don't need a college degree in climate science to know this, do you? Clouds block a large portion of the sun's heat energy. They also trap some energy from returning back into space, such as at night time when the land or water is still warm from solar heating during the prior day.

Simple reality check: on a typical day, do you think more than 0.04% of our planet's sky has clouds in it?

I could ask you to pick a ballpark number as a wild guess. But this is silly. Clouds have a HUGE impact on the amount of energy coming IN and OUT of our planet. And we all know they are ORDERS OF MAGNITUDE more than 0.04% of our sky.

Want an actual scientific number? It turns out a satellite with the acronym MODIS was launched in 1999 with one goal of measuring the average cloud cover of the planet. If you care about the super-geeky technical name for MODIS go to its Wikipedia page. The result, as described in a technical paper published by the IEEE technical society is:

67% !!!

Tell me again why we think CO_2 justifies a global crusade?

To their credit, modern climate models do include clouds and cloud effects. But can anyone – especially a genuine climate scientist – tell us with any credibility that taking action against 0.0015% of our atmosphere makes any sense compared to these clouds that are clearly overwhelmingly more influential? And more variable!!

In fact, a modest but dedicated cloud-seeding program would have an impact NEXT WEEK on global temperatures if we really cared to do anything effective about it. Assuming the increasing trend and the negative consequences are real, of course.

The fact that nobody credible is proposing this (or it hasn't gotten a consensus of support) proves that something very wrong and dangerous is going on instead.

If it isn't the heat capacity of CO_2, or the radiation qualities of CO_2, what could possibly make CO_2 so special??

Instead of paying people to run those massive computer models that are trying to analyze and predict the effects of 0.039% of the atmosphere, maybe we should spend a few dollars to look at—*and explain*—the basic physics!

DON'T FORGET THE FULL CONTEXT

Don't forget this fundamental fact too: whatever characteristic makes CO2 so important, it is still **only 0.04% of the atmosphere.**

And the **human-created portion is only 4% of *that*.**

What quality of CO2 could possibly be so unique and powerful that it justifies the current crusade against **0.0015% of the planet's atmosphere??**

Then there is the inconvenient fact that the countries of China and India, each with over 1 billion people, are contributing an increasing amount of the human output of CO2 every year. And they have no intent of slowing down.

WHO HAS THE BURDEN OF PROOF

Until we have a basic explanation of the facts and physics, there is nothing wrong with being skeptical about the current climate change hysteria. **The burden of proof rests with those who are making the claims of a crisis that demands action.**

Arguments from authority, from intimidation, from the alleged consensus, begging the question, or any of the other bogus logical fallacies that dominate the current debate are not going to cut it anymore. It's time for a reality check!

THE REAL DANGERS OF THE CLIMATE CHANGE CRUSADE

Instead of spending money, time, and brainpower on ways to slow down or restrict human activity and prosperity, why not approach any potential problem with an attitude of abundance and optimism?

We can spend billions or trillions of dollars trying to stop the sea level from rising a few inches (even feet) over the centuries. Or we can invent new ways to adapt and live happily in a changing environment. Maybe design and build dams and levees that work! Or here's a radical thought: let people make their own free choices on where they want to live and move. And when!

We have plenty of examples throughout history that show it is very dangerous to try and control the mass behavior of millions of people. That's a genuine danger that has led to the deaths of **tens of millions of people** in the last hundred years.

If you really want to be rational and use scientific probabilities, **the biggest danger to human life ending painfully or tragically prematurely on this planet comes from other humans.**

It will not come from natural disasters or geologic changes!

The debate about this controversy has evolved into another real danger: political oppression.

Why are so many people afraid of approaching this problem from a perspective of persuasion and free discourse? There are serious proposals from politicians and policy makers to make "questioning climate change threat" a crime punishable by fine—or worse!

When there is no immediate threat to anyone's life—as in RIGHT NOW—isn't it dangerous to call for the elimination of free speech? Public shaming? Simply for asking some honest but unpopular questions?

IF WE MUST DO *SOMETHING*…

If we WERE to do anything…it should be leaving people free to pursue, research, discuss, collaborate, invent, and earn a living (and/or profit) from whatever efforts they decide is best. If you still think there are threats or opportunities worth pursuing, as long as you don't compel me to pay for it or organize politicians to start taxing me for simply being alive (because you and I are both putting CO_2 into the atmosphere right now), I'm not going to stop you.

Maybe you'll actually create some device, process, or energy source that is more efficient and clever than what we have today. That would be a great (and real) improvement, which I would be happy to see in the world. Until then, keep the basic facts of reality in mind, please.

And remember a sense of proportionality as you debate and decide what we all should do on our amazing planet!

That's all I have to say about the physics of AGW or manmade global warming and climate change. I hope it was worth your time and effort to read it.

STILL OPEN TO DEBATE AND LEARNING

Care to challenge or correct any of these premises, assumptions, facts, or calculations? If you are operating from a philosophy that respects facts, an objective reality, and a rational sense of proportionality, I'm willing to hear from you. Please go to the Amazon page for the book and leave a comment there.

If you would rather revert back to arguments from authority or fear or some other bogus premise...you can do that if you want but I'm just going to reply to people who stick to the facts and the physics about our atmosphere and the temperature mechanisms within it.

Plus, remember that whatever you post on Amazon's page is going to be there for everyone to see (and flag if it's out of line).

Do I really think this one book will settle the climate change debate? No, of course not. But if you think more people should read this, I'll be humbled and very grateful if you'll find it worth recommending or sharing. Leaving a review online is wonderful too! However you choose to tell more people about it (whether you are happy you read it or you are gaining assistance to prove it wrong), you are doing the world a favor.

Thanks again for reading this book. It's my humble attempt at contributing some facts, logic, and proportionality to the debate that is consuming so much of our world today.

WHO IS CALVIN FRAY?

Calvin Fray is someone who cares about the truth, rational thought, and doing what is best—for each us as individuals and, as a consequence, for all of us as a species. He prefers for his words and actions to speak for themselves—to be so honest, virtuous, and effective that it doesn't really matter who Calvin Fray is beyond that.

Has he been successful with this book? That is for you to be the judge, as with everything else in your life.

Do you still want to know more about Calvin Fray? He prefers that you don't spend any more time and effort thinking about him. It's probably a pen name anyway…

Instead, please think about others you know who would benefit from reading this book. Tell your friends, family, and colleagues about it—whether it's because you agree with what is here or you want help to prove it is wrong.

Then, if you still want to learn more about how and why someone would write a book like this—and why it matters—the better question to ask is,

Who is John Galt?